The Correlation Of Occult

Teachings With The Findings Of

Academic Science

Khei

Kessinger Publishing's Rare Reprints

Thousands of Scarce and Hard-to-Find Books on These and other Subjects!

- Americana
- Ancient Mysteries
- Animals
- Anthropology
- Architecture
- Arts
- Astrology
- Bibliographies
- Biographies & Memoirs
- Body, Mind & Spirit
- Business & Investing
- Children & Young Adult
- Collectibles
- Comparative Religions
- Crafts & Hobbies
- Earth Sciences
- Education
- Ephemera
- Fiction
- Folklore
- Geography
- Health & Diet
- History
- Hobbies & Leisure
- Humor
- Illustrated Books
- Language & Culture
- Law
- Life Sciences
- Literature
- Medicine & Pharmacy
- Metaphysical
- Music
- Mystery & Crime
- Mythology
- Natural History
- Outdoor & Nature
- Philosophy
- Poetry
- Political Science
- Science
- Psychiatry & Psychology
- Reference
- Religion & Spiritualism
- Rhetoric
- Sacred Books
- Science Fiction
- Science & Technology
- Self-Help
- Social Sciences
- Symbolism
- Theatre & Drama
- Theology
- Travel & Explorations
- War & Military
- Women
- Yoga
- *Plus Much More!*

We kindly invite you to view our catalog list at:
http://www.kessinger.net

INSTRUCTION VI.

THE NEBULAR HYPOTHESIS.

THE CORRELATION OF OCCULT TEACHINGS WITH THE FINDINGS OF ACADEMIC SCIENCE.

Occult and Physical Science agree.—In Instruction Number One of this series, the statement is made that "Occult science agrees with Physical Science in the application of the Nebular Hypothesis to the Creative Scheme, insisting, however, upon the direction of the nebulic activities by the wisdom and intelligence of the Hierarchies described."

It is therefore important that the student at this point in the course, be accurately informed as to just what the Nebular Hypothesis is, in detail.

First, we will take up the various definitions:

What a Nebula is.—A NEBULA is one of the masses of gaseous matter found in different portions of the Heavens.[1]

Their quantity.—Nebulae are exceedingly numerous, over 12,000 being easily within the range of vision of the three-foot mirror telescope, and many times this number may be seen by the new five-foot reflectors recently constructed.[2]

How named.—Nebulae are named either after their color and shape, their discoverer, or their position in the heavens. For instance—

Examples.—a Annular or Ring Nebulae—Dusky in the center and bordered by a lighter ring of light.

b Cometary—Round, with star-like nucleus in center, resembling the average comet.

c Crab—So-called by Lord Rosse, on account of the claw-like appendages.

d Dumb-bell—The luminous cloud of star-dust of gas in the constellation of the Fox, and resembling a Dumb-bell.

e Fish-mouthed—From the shape of one seen in Orion.

f   Green—Due to the greenish color of some.

g   Hind's Variable—Last seen in Taurus and no longer visible.

h   Horseshoe, Omega or Swan—Seventeenth in Messier's list; so named from its shape.

i   Keyhole—A nebula of that shape in the constellation Argo.

j   Andromeda—in the girdle of Andromeda (Cons.)

k   Orion—The largest known; in Orion's sword hilt.

l   Owl—A nebula in Ursus Major, shown in old maps as an owl's head.

m   Spiral—From their appearance.

n   Stella—A body of numerous distant stars appearing *like* a nebula.

o   Variable—Varying in brightness. Hind's and Struve's.

p   Whirlpool—In Canes Venatici, so named by Lord Rosse, from its remarkable spiral formation.

q   White—One whose continuous spectrum does not show the bright lines that ordinarily characterize that of a nebula.

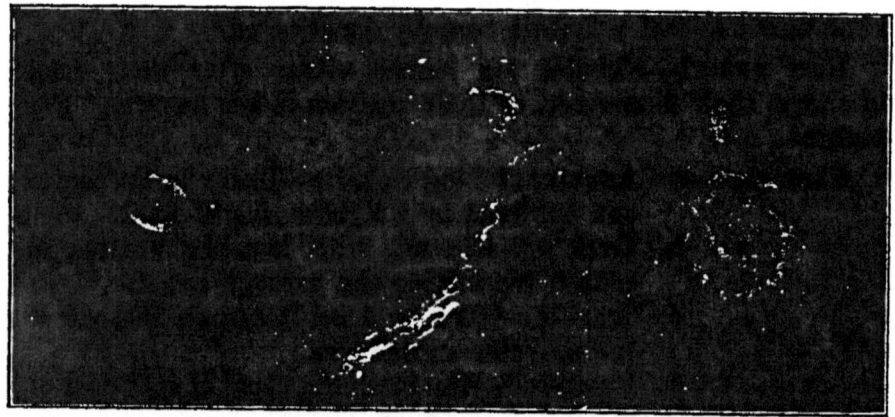

FIG. 19

EXAMPLES OF NEBULAE. LEFT, RING NEBULA IN LYRA:
CENTER, NETWORK NEBULA IN CYGNUS: RIGHT, SPIRAL
NEBULA CANUM VENATICORUM.

—*(Photos by Mount Wilson Solar Observatory)*

What the Nebular Hypothesis is.—The Nebular Hypothesis is that the Solar System existed originally in the form of a nebula, which by cooling, condensing and revolving was formed into the sun and rings of matter, which later were consolidated into planetary bodies; applied also to all the heavenly bodies.[3]

. **Who Formulated and Developed it.**—"It was indefinitely conceived by Swedenborg, more clearly outlined by Kant, Herschel, and Laplace, expanded as one of the general processes of evolution by Herbert Spencer, brought down to the basis of a physical theory by George Howard Darwin, and amplified to date by Thomas Crowder Chamberlain."[4]

What Draper says.—(Intell. Develo. Europe, Vol. ii, p. 281, 1878). "The Nebular Hypothesis compels us to admit that all the ponderable material now existing as constituting the various bodies of the solar system once existed and extended in a rarefied or nebulous and rotating condition beyond the confines of the most distant planet."[5]

The overwhelming preponderance of the nebulae are spiral in shape, and constitute the typical form of the class.

Nebulae may be technically defined as luminous whirlpools of matter. The central luminous nucleus is a gaseous, incandescent body like our own sun, with two spirals leading out from it, having irregular foci of light, with filmy veils of luminous smoke. This latter is matter that has broken forth from the central body.[6]

The extent or span of a nebula is enormous, often to be measured only by unthinkable millions of miles. For instance, there is a nebula in Andromeda that is so wide that light, which travels at 186,000 miles per second, requires eight years to span it. In another way, its measurement may be given at five hundred thousand times 93,000,000 miles, which last is the distance of the earth from the sun.[7]

These nebulae are the matrices from which solar systems are born.[8]

The old teachings of physical science were to the effect that our solar system originated from a super-heated gaseous globe, which contracted as it cooled, and from time to time threw off from its equatorial belt substance that later became planets.

Two hypotheses.—The later findings of physical science resolve themselves into two distinct hypotheses, one of which, the second to be described, coincides precisely with the teachings of the Brotherhood.

Fig. 20

The Great Nebula in Orion. One of the most wonderful sights ever revealed by the telescope. It is believed by some scientists to be the birth of an universe. Its size can only be estimated by imagining a sphere with a diameter as great as that of the earth's path around the sun, and then multiplying it by a million. It illustrates the processes that brought the present celestial systems into being.

—(Photo by Yerkes Observatory.)

First hypothesis.—According to this concept, a spiral nebula is a system of worlds in the making. The central nucleus is to be the future sun. The luminous spots that lie irregularly within and along its spiral arms are the nuclei of future planets, having been thrown off from the central nucleus as it progressed in movement.[9]

Planetesimals.—These luminous spots that are destined to become future planets are termed "PLANETESIMALS."[10]

Entire nebula moves.—The entire nebula moves in a complete mass, and as time passes the growing, smaller nuclei draw to, or attract through gravitation, matter from the surrounding mass until each revolves in a clear space.

Original nebula becomes the Sun.—The original central nucleus draws to itself the myriads of smaller and minor planetesimals in its neighborhood and in time becomes the detached central sun.

Second Hypothesis, accretion to mass.—The second aspect of the nebular hypothesis follows that just given, with the addition of taking into account the accretion to the planetesimals of the matter built up around them by perpetual meteoric bombardment, gathering matter more rapidly through gravitation as their mass increases, and from the beginning developing a high interior temperature through impact and compression. This growth is continuous, as for instance, our earth sweeps up something like a hundred million meteors each day, these meteors being belated fragments of the original nebula.[11]

Compare this with the teaching that the creating God occupies a definite location in so-called space which he fills with his aura, and differentiates out of the surrounding cosmic root substance matter which he energizes to a different status of vibration, etc., and you will find that although differing in verbiage, the process is the same, except that Rosicrucianism insists upon the intelligent ordering and direction of the activities and movements for definite purposes.

Origin of nebulae.—As to the origin of the spiral nebulae there are also two widely differing theories in physical science.

1st That hundreds of millions of years prior to its nebulic formation, a nebula was originally a gaseous star which in its progress entered the neighborhood of another star.

Roche's Law.—Now according to Roche's Law of Limits, the critical distance of 2.44 radii is sufficient to cause the force of gravitation to tear asunder the structure of both bodies, causing both globes to explode like bombs and their fragments of comminuted world stuff to be scattered out into space.[12]

Teaching of Arrhenius.—But here again the greater weight of scientific authority supports unintentionally the Brotherhood teachings, for the savant ARRHENIUS voices the consensus of opinion that the spiral nebulae are formed from the universal "Cosmic Dust," (Rosicrucian Cosmic Root Substance, the 2nd and real theory).[13]

Nebulae formed of "cosmical dust."—According to this teaching, the nebulae are formed of this "cosmical dust" under the influence of the pressure of "radiant energy" (the activity of the creating God). The electrified particles of world stuff from other suns collide in space and form meteorites which later aggregate into the larger mass under the influence of gravitation.[14]

How they become luminous.—Such nebulae are cold to the point of absolute zero, and become luminous through the impact of the electrified particles and glow like the rarefied matter exhibited in the phenomena of the vacuum tube under electrical influence.[15]

How a star is born. Cold at first. Its heat comes from within, not from the Sun.—A star, then, is a body born out of the cosmic mist of a nebula. Although cold at first, even though born from an incandescent nucleus, it begins to give off intense heat as it becomes hotter and hotter through its contraction. It ultimately attains the supremest white brilliance as illustrated by Sirius and Procyon. Spectral analysis at this stage reveals the light gases helium and hydrogen.[16]

Becomes yellow.—As it cools later in its progress it becomes yellow in color and shows evidences of calcium, iron, etc. Our own sun and Arcturus are in this stage.

Then red.—Still later it becomes red, and the spectrum indicates larger evidences of carbon. Such are Betelgeuse and Mira. These processes require billions of years.

Rejuvenation.—When completely cold, a star can be brought again into light and activity by collision, by which new energy and momentum are imparted, and so the cyclic process goes on indefinitely, and will go on until all the matter in the universe has been aggregated into a single mass.

Cosmic Night.—The periods of darkness of each star are the Cosmic Nights already referred to, intervening between each Day of Manifestation, and the activity of the creating God is exhibited in the dissemination of the "Cosmical Dust" and its reassembling into nebulae for recreative or new creative processes.

Light.—Throughout occult science, the principle or factor of LIGHT receives unusual emphasis. And well it may, for in both occult and physical science LIGHT will be found to be the active principle in creative activity, the same principle which first came into being, at the primal fiat, "let there be LIGHT," and after there "Was LIGHT" the rest of the creative processes continued. This is shown in physical science by the "RADIATION PRES-SURE OF LIGHT." It being understood that the planets had their origin not as gaseous rings but as electric bombardments from the original nucleus, we can study the action of Radiant Light pressure in seeing how this was brought to pass.

Recapitulation.—First, let us recapitulate. Our earth, for instance, never was either a gaseous ring or a liquid globe as liquids are ordinarily understood by laymen.[17]

How a star is "cold" at first.—Second, we have stated that a star born out of the cosmic mist of a nebula was "cold at first." This apparent paradox is easily explained when we understand that after being thrown off from the parent nucleus for a time its substance is widely distributed and semi-diffused throughout a part of the nebulic mass, and therefore cools, in its atomic structure, until assuming the attributes of its ultimate shape it begins to generate heat itself by the process of contraction.

Therefore we may state at this point the following affirmation:

AUTHORITATIVE ROSICRUCIAN AND ACADEMIC SCIENCE TEACHES THAT THE TRUE NEBULAR HYPOTHESIS OR EXEGESIS IS—

1st That the central nucleus of the nebula is the future sun of a new solar system. Rosicrucianism amplifies this affirmation by stating this nucleus to be the central focal point of the activities of the creating God who may be operating through the nebula under observation.

Planetesimals.—2nd That the bodies thrown off from the central nucleus, namely, the smaller nuclei of varying densities and proportions, are the planetesimals, or future planets of the solar-system-in-the-making.

Not rings but built up.—3rd That these planetesimals are not rings but are thrown off in mass from the central nucleus, and that each planetesimal or lesser nucleus builds up by attraction and meteoric bombardment.

How axial motion and spherical shape are obtained.—4th That the individual nucleus or planetesimal derives its initial axial motion from the frictional contact with the nebulic structure, which is in constant interior motion, producing also the spherical shape.

FIG. 21

One of the anomalies of our solar system. The Planetoid "EROS." Planets are considered to be more or less spherical in shape. Eros is an exception. It is practically a huge mountain in space, "without form and void," and as it turns upon its axis first one corner and then another is presented to view. It has not sufficient gravity to draw its structure into symmetry, and remains as when launched into space. It tantalizes astronomers to know that Eros passed very close to earth January 24, 1894— before it was recognized, and that so near an approach will not occur again until 1975. Jupiter, the ponderous planet, is usually regarded as a "benefic" yet it is really the most troublesome of all of Sol's family, for it appropriates from the smaller planetoids and comets about what it desires and many astronomers believe that Eros is the remains of a planet which has suffered at the hands of this "thief of the skies."

Origin of orbital motion.—5th That each individual nucleus or planetesimal derives its orbital motion from the spiral sweep of the entire nebulic mass, and that as the entire nebula moves as a unit carrying all its planetesimals with it, so after ultimate clearing of the nebulic structure, the resultant solar system continues to move in a general orbital sweep, carrying its planetary family as a unit.

Origin of nebulae.—6th That the nebula has its origin in "Cosmic Dust," Rosicrucian Cosmic Root Substance, the product not of the collision of worlds or planets but of the electrified particles of world-stuff from other suns (which are the physical vehicles of the creating Gods of those particular systems), which collide in the space indicated by the new-creating God, forming the meteoric masses which later aggregate into the larger nebulic mass under the influence of gravitation.

Direction of our solar system.—Our entire solar system moves forward in space at the rate of 12½ miles per second, in the direction of 15 degrees southwest of Vega, in the constellation Lyra. Vega is also approaching us at a rate that brings the two systems nearer by 2,000,000 miles each 24 hours.[18]

God Is Light, Life and Love.—Rosicrucian teachings affirm that the triune Deity is Light, Life and Love, and it is a fact that wherever the last two are found, the first will always be present, exoterically and esoterically. Sometimes chemical methods must be employed to demonstrate its presence exoterically, nevertheless it will always be found.

Light, great cosmic force.—Therefore, LIGHT may truly be said to be one of the greatest of Cosmic forces manifesting the Will of the Absolute.

Radiation Pressure.—"Radiation Pressure" means the pressure and activity of light. It is the direct force that preserves the integrity of the nebula that is destined to be a future solar system; preventing, by overcoming for the time being the force of gravitation, the fine particles of nebulic matter from gravitating toward other interspatial bodies.[19]

Reduced to a practical affirmation, then, Rosicrucian teachings agree with Professor Campbell, who puts it thus:

Nebulae and Light.—"A NEBULA CONSISTS WHOLLY OR IN PART OF FINELY DIVIDED PARTICLES OF MATTER WHICH ARE THRUST HITHER AND YON BY THE LIGHT PRESSURE OF MYRIADS OF INCANDESCENT STARS, IN SEEMING DEFIANCE OF THE LAWS OF GRAVITATION. IN DUE COURSE, HOWEVER, THE FINE PARTICLES OF MATTER BECOME AGGREGATED AND THUS BECOME TOO LARGE FOR THE LIGHT WAVES TO ACT ON THEM EFFECTIVELY."[20]

Succumbs to Gravitation.—This aggregated matter then becomes so concentrated as to form the more or less solid body that we call a star. "Thenceforth, this body, undergoing a series of transformations which cause it to be difinitely labelled, must move in response to the aggregate pull gravitationally of the stellar bodies that make up the universe."[21]

Comets.—Many uninformed persons mistake comets for nebulae and regard the two as identical. This of course is wrong, and some erroneously believe that comets exercise a baneful influence upon the earth, some occultists even going so far as to teach that planetary continental modifications are actually produced by them. Nothing could be farther from the truth. Over 650 comets are

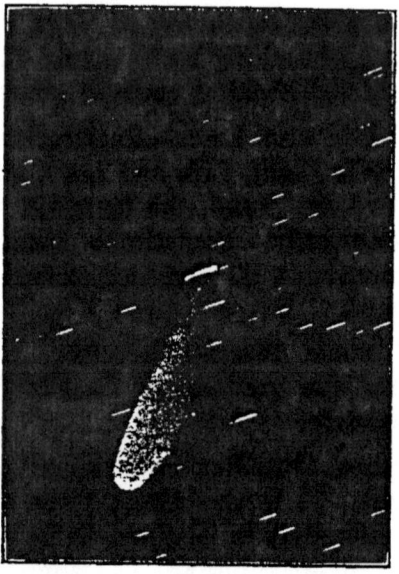

FIG. 22
A GOOD EXAMPLE OF A COMET.

now recorded and classified as Parabolic, Elliptical and Hyperbolic; comets of Long and Short Duration. Comets move for the most part, however, in parabolas, and their orbits have no evident relation to the plane of planetary motions. This fact indicates that they need not be considered as belonging in any true sense to the Solar System itself, but are merely visitors therein from interstellar spatial regions.[22]

Their Parts.—A comet consists of the coma, or shining matter, the nucleus, a bright point near the center of the coma, and the tail or streamer. This tail is not discarded matter, as is shown by the fact that while it *follows* the coma as the comet *approaches* the sun, it precedes the coma as the latter *recedes from* the sun. The tail is always *away from* the sun.

Size.—The size of a comet is almost incredible, in general the head or coma being from 40,000 to 100,000 miles in diameter; the comet of 1811 having a head 1,200,000 miles in diameter. The tail of the great comet of 1882 was 100,000,000 miles in length. The head contracts as the comet approaches the sun.

Mass and Density.—Yet a comet's mass is insignificant, and its density inconceivably small, "much below the density of the residual gas left in the best vacuum science can produce."[23] Comets are not self-luminous, but shine by reflected light, as is proven by their relative luminosity as they approach or recede from the sun.

No evil effect upon the earth.—Instead of exerting a destructive effect upon our Earth, when the planet passed through a comet some years ago, the only noticeable effect was a general luminescence or almost phosphorescent effect, dimly perceived. On the other hand, when a comet comes into contact with another planet or enters that planet's orbit, it is apt to be either broken up or "made captive," as is often done by the ponderous planet Jupiter, sometimes called the "thief of the skies." A comet has been caught by the attractive power of this mighty body, made captive and compelled to move an orbital prisoner.

Meteorites.—The same fear expressed relative to comets has also been manifested in regard to meteorites; broken bits of other worlds and disintegrated worlds. Over twenty millions of these meteorites enter our atmosphere every twenty-four hours, yet of all this great number only about 700 have actually reached the earth's surface, the remainder being dissipated or consumed in our atmosphere by friction created through their own inconceivable speed. The ocean floor is said to be covered with a thin layer of the ash.

Shape of Our Universe.—So far as can be observed, the "bulk of the stars, exclusive of those of the Milky Way, form a vast lens-shaped structure, and as we attempt to picture in the imagination this vast lenticular structure, comprising in the aggregate all the matter in the universe, the thought comes naturally to mind that the entire system, with its hundred million or thousand million stars, may be whirling about the axis of the galactic poles, with some giant sun, so distant that it seems to us no different from other stars—at its center of revolution."[24]

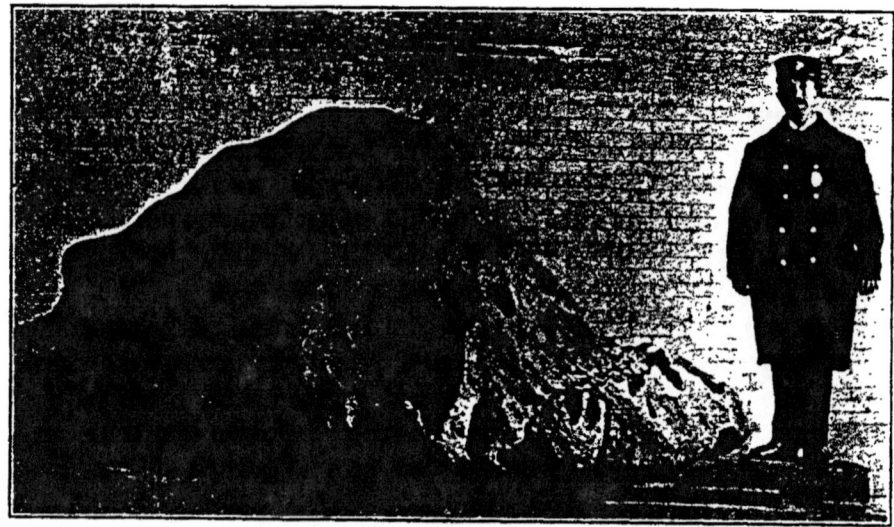

FIG. 23
THE FAMOUS WILLAMETTE METEORITE. WILLAMETTE
VALLEY, NEAR OREGON CITY, OREGON.
—(Courtesy of the American Museum of Natural History, N. Y.)

The Via Lactea.—The Milky Way or Via Lactea mentioned above, may be described as the luminous belt surrounding the heavens in a great circle. It varies both in width and brightness, and for a third of its extent, that is, from Cygnus to Scorpio, it appears divided into two parallel streams. The Via Lactea is more commonly known as the "Galaxy," and is made up almost wholly of stars of the eighth magnitude and less. It contains many true star-clusters, but few real nebulae, and in some places the stars are too thickly clustered to permit of accurate record or estimate.

In early times the Milky Way was the origin of much really important mythos and deific personification; in mediaeval times it was the subject of a vast deal of superstition; in modern times it is the subject of critical investigation. In its entirety, it exercises an undoubted influence of an occult nature, not only upon our planet, but on all others within a reasonable radius, but its influence is that of a benefic, and by the ancients it was supposed that the Milky Way exerted a pull that counteracted the force that held all bodies to the earth's surface. The ancients believed that but for this benefic influence, we should be inevitably drawn or "sucked into" the earth's crust. Needless to say, this idea is not entertained as other than mythos. The influence exerted for our benefit is manifested through the combined forces of the enormous number of its starry members, concentrated upon the material forms utilized by organic life upon our planet, and preventing too great consolidation or crystallization and operative principally through our Etheric and Astral vehicles.

Thus from the initial activity of Light, exerted as Radiation Pressure, results the mighty aggregation of a solar sun and its attendant planetary family; a slight idea of the immensities involved being indicated when we remember that the actual mass of our sun is 332,000 times that of our earth,[25] and the latter weighs approximately 5,272,600,000,000,000,000,000 tons. (It is interesting to note in comparison the weight of the Great Pyramid, 5,272,600 tons.) Yet the whole bulk of this little sun, chief luminary to earth, a fifth rate body in a minor solar system, is gaseous; although, due to enormous gravitational pressure, the sun's interior structure has a consistency more like dense liquid than what we ordinarily conceive of as gas, which explains our previous statement.[26]

All nebulic nuclei pass through same process.—All the nuclei pass through the same evolutionary processes and stages as does the parent nucleus or sun. Thus each in turn has its gaseous, liqueous (gravitational compression of the gaseous) and ultimate solid status. Jupiter, for instance, is 1/3 denser than water, and Saturn 3/4 as dense.[27]

Light and the Radiometer.—It is to Light that we owe our knowledge of the constituency of our planetary neighbors, for through the spectrum analysis, it reveals the presence of all known and some unknown elements and metals. Light rays both above and below the known spectrum are now powerful agents in modern use, such as the X-Rays, Ultra Violet, Helium and other rays, and for the simplest of all illustrations, the student has only to purchase an inexpensive radiometer at any optical store and watch the action resultant upon the impinging of the light rays or radiant energy upon the delicate apparatus. The blackened sides absorb, and the bright sides reflect. Some of these instruments are so delicate that a candle placed more than 1/3 of a mile away will turn the vanes through nearly 100 scale divisions.[28]

Human light rays.—And last of all, the energy of light is shown by its perceptible radiation from the homo or human being under proper circumstances.

Light acts only on infinitesimals.—Now, powerful as light has been shown to be, it must be noted that it is capable of acting only upon infinitesimal particles, and that when such particles aggregate, they enter the domain of gravitation. Prof. Arrhenius has estimated the size of a particle of matter that may be driven before a light wave, but it is indescribable in figures of comprehensible measurement.

Radiant Energy.—Radiant energy requires a medium of transmission, and between the nebula as it comes to our vision and the "cosmic dust" or Cosmic Root Substance from which it originated lies a vast gulf. Therein lies the whole range of matter from Mass, Molecule, Atom or Ion, Electron or Corpuscle, and Ether.

We will define them in order.

1 **Mass** is an aggregation of molecules. It is distinct from weight with which it is often confounded, weight being simply an indication of mass.[29]

2 **A Molecule** is the smallest particle of a substance that can exist in a free state, and which has the same composition as any large mass of the substance.[30]

3 **An Atom** is the smallest particle of an Element that exists in any molecule.[31]

4 **An Ion** is an atom or radical of a substance resulting from electrolytic decomposition or dissociation. Moving in the direction of the anode or positive pole, it is called an electro-negative ion. Moving in the opposite direction, it is the electro-positive ion. Gaseous ions are combinations of molecules with an electron, and are electro-conductors.[32]

Broadly speaking, then, an Ion is an atom charged with electricity.

5 **A Corpuscle** is an electro-negative ion.

6 **An Electron** is identical with the corpuscle, but it is the term used in lieu of corpuscles when explaining the ELECTRONIC THEORY OF MATTER, which is that "all mass is the mass of the ether, all momentum, whether electrical or mechanical, the momentum of the ether, and all kinetic energy the kinetic energy of the ether.[33]

All of these definitions will be amplified at the conclusion of this Instruction.

7   **Ether**   is the great mystery of physical science. It
                     is impossible to give to this mystery any
                     "simple" explanation.

A medium.—ETHER may be said to be the imponderable,
hypothetical, but none the less real, medium; tenuous and elastic,
diffused throughout all Cosmos and which forms the avenue of
transmission of radiant energy. It cannot be confined, and therefore much must long remain a matter of hypothesis regarding it.[34]

Its family place.—We do, however, know positively somewhat
of its attributes and properties. Ether is not to be considered
non-matter as such. Instead it is the simplest and lightest of the
elements, an inactive GAS of the ARGON family, being assigned
to position x of the Zero group. The atomic weight of Ether has
been ascertained to be one millionth of that of the hydrogen atom,
thus permitting its atoms to maintain a tremendous velocity,
which explains its interpenetrating and all pervading presence.[35]

Rosicrucian concept of Ether.—As a matter of fact, however,
Ether is of surpassing importance to the Rosicrucian, for it is to
him the link between Spirit and Matter, Divinity and Humanity.
We are taught that "In Him we live and move and have our being." We know that in the Ether are fulfilled all these conditions.
This does not imply the reduction of Deity to a gas, but it must
be kept in mind that when we speak of gases we mean the final
disintegration from solids and liquids into the vaporous, ethereal
state that, in comparison with our physical world and its three
dimensional concepts, must be largely analogous to the contrasting
term "gaseous."

Ether, the densest substance known.—Now although Ether
presents a gaseous and even fluidic state, comparatively speaking,
offering no resistance to the movements of objects through it, acting as a medium of transmission of waves of energy, penetrating
and permeating all things in cosmos, the very attributes of omni-
science and omnipresence, yet IT IS THE DENSEST AND MOST
MASSIVE STRUCTURE IN THE COSMOS—again like omni-
science in this attribute.

Matter, least substantial.—Rosicrucianism has always taught
that the material world is the reflection of the spiritual world.
Physical science now asserts that WHAT WE CALL MATTER IS
THE MOST EPHEMERAL AND LEAST SUBSTANTIAL THING
IN THE UNIVERSE.

Ether.—The unit particle of electricity which we call a corpuscle or electron owes its mass entirely to an infinitesimal quantity of Ether which is bound up with its substance.[36]

Its density.—The density oi ether attached to such a corpuscle has been found to be 2,000,000,000 times the density of lead.[37]

It may seem inconceivable that we are immersed and exist in a medium two billion times as dense as lead,[38] but the latest findings of physical science explain this by the statement that

"MATTER IS COMPOSED MAINLY OF HOLES" to such an extent that the volume of ether disturbed by the movement of matter through it is infinitesimal compared with the volume enclosed by it.[39]

Size of Ether granules.—Ether is composed of spherical granules so infinitely small that 700,000,000,000 of them in one line could lie in the trough of an ultra violet wave, which is one seven-thousandth of an inch.[40]

Where Rosicrucians and Physicists agree.—Here again Rosicrucians and academic physicists contact, for Prof. Reynolds says, "these granules are THE ULTIMATE OR PRIMORDIAL ATOMS, perfectly spherical and perfectly rigid, infinitely small in comparison with the electron or corpuscle.[41]

Ether granules are Cosmic Root Substance.—And Rosicrucian philosophy asserts that Ether is composed of PRIMORDIAL ATOMS OF COSMIC ROOT SUBSTANCE, the "Cosmic Dust" of which the nebulae or worlds-in-the-making, are formed.

Matter likened to bubbles.—Prof. Reynolds further likens matter to bubbles. Prof. Mackenzie explains this as follows: "You have all seen bubbles moving in water. Reynolds shows that the earth and all other material bodies move through space in a similar manner. They are less dense than the medium in which they exist, and their movements are due to differences of pressure in the surrounding medium (Cosmically, the pressure due to the activities of the creating God). Real mass is not in the thing materially which we see, but in space where the eye sees nothing. The sober conclusion of the most advanced Dynamical Science is that MATTER IS A NEGATIVE THING SO FAR AS ITS MASS IS CONCERNED, AND THAT THE SPACE OCCUPIED BY MATTER CONTAINS VERY MUCH LESS THAN THE SPACE WHERE NO MATTER EXISTS."[42]

What the Universe consists of.—"The entire universe of matter consists essentially of little maladjustments or flaws in the universal granular ether"[43] and note this, Rosicrucian students, that each SPHERICAL GRANULE IS ASSOCIATED WITH TWELVE OTHER SPHERICAL GRANULES THROUGHOUT THE ENTIRE ETHERIC MASS.[44]

Nature a series of reproductions.—All Nature is a series of reproductions of processes. As the nebula originates the future solar system with its orbital processes, so in the smallest ponderable structure we find the same analogy.

The Creative Link.—The electron, as we have seen, is the electro-negative corpuscle bound up with the infinitesimal portion of Ether. This latter, the Ether, may be considered Ether in its free state and therefore the positive element or pole, while the corpuscle or electron may be considered the crystallized Ether. Herein is the creative link.

Not crystallized in the sense of greater density than the primordial Ether, but in the sense of a focus or concentration. Thus we have in sequence—

Descent of Cosmic Root Substance.—1 Cosmic Root substance, the spherical granules of ether.*

2 Concentrated granules or corpuscles (electrons), negative.

3 Free granules, positive (Essential activity of Deity). The two combine to produce the

4 Positive and negative ions, which compose the

5 Atoms, which combine to produce

6 Molecules, which combine to produce the

7 Mass, which is of two kinds,

8 Elements, an element being a substance whose molecule contains only one KIND of atom—and

9 Compounds, a compound being a substance whose molecules contain two or more kinds of atoms, expressing first the invisible, tenuous, spiritual state, then known as

---

*Another term has been added to the definition of the component parts of the atom. It is offered by Dr. Irving Langmuir in his new theory of the structure of matter presented at the annual gathering of the National Academy of Sciences early in 1920. Speaking of the sub-divisions of the atom he introduced the term, "Quantel" which, he said, consisted of two parts, positive and negative, present everywhere in space, moving in all directions with the velocity of light and capable of passing through matter. They constitute, he said, what has heretofore been known as the "ether of space." As an instance of the extent to which the materialistic concept of abstract propositions may be carried, Prof. Langmuir is said to have asserted that *"space and time have a structure analogous to that of matter."*

10 Invisible matter, next by further concentration and crystallization into the state of

11 Visible matter, or the physical world, ultimately again returning to the

12 Invisible or spiritual world, ascending spirally toward its

13 Creative source—DEITY.

All things "come from the center," travel around the circumference of evolutionary processes, upward, and return again to the center, as science bears out by its affirmation that ultimately, in the words of Profs. Stewart and Tait, it is certain "that age after age the possibility of such transformations (of energy) is becoming less and less; and, so as far as we yet know, the FINAL STATE OF THE PRESENT UNIVERSE MUST BE AN AGGREGATION INTO ONE MASS OF ALL THE MATTER IT CONTAINS, i. e.—the potential energy gone, and a practically useless state of kinetic energy prevailing, i. e.—uniform temperature existing throughout the mass."[45]

Action of the 12 around the 13th.—Everywhere the activity of the twelve around the thirteenth will be found, for in the structure of the Atom we find it composed of electrons making up definite planetary systems within, circling about with infinite speeds in regular orbits, and one electron dislodged from its atomic system would dash from one atom to another at the rate of 40,000,000 times per second.[46]

Students will recall that the attributes of the second aspect of triune Deity the Supreme Being, were the Word, POWER AND MOTION; and of the third aspect, WILL, Wisdom and ACTIVITY.

How Deific attributes manifest.—Rosicrucian philosophy teaches that the Power and Motion of the Supreme Being, and the Will and Activity of the Solar God, are the dualities that express themselves respectively as the free Ether and the corpuscles that form the basis of all later development in the cosmic scale.

Ether a vehicle.—The creative WORD and WISDOM is transmitted via the Radiant Energy of which Ether is the vehicle.

Energy a ray from Deity.—A ray from Deity, a wave or ray of Radiant Energy dissociated, develops heat, light and ultra violet, the prime factors in creative process.[47]

Etheric equation.—Ether also represents mass to the Cosmic Root Substance granule, as does physical mass the agglomeration and aggregation of structural physical atoms.

Number of Elements.—An element was defined as a substance whose molecule contained only one kind of atom. There are seventy (the mystic number) such elements known to science.

Their combinations and permutations.—These seventy atoms or elements may combine in a number of permutations reaching 250,000. Thus they furnish vehicles for the activities of all demonstrable forms of life.

Ether furnishes Spiritual Spheres.—So too in the Ether, the corpuscular combinations furnish sublimate states or "spiritual spheres and planes" for the activities of those forms of spiritual life more closely approximating the celestial than the terrestrial.

The common triad.—Therefore, in common with physical science, Rosicrucianism holds with equal importance the knowledge of the triad

     1 Matter   2 Ether   3 Energy[48]

as essential to the logical understanding of spiritual worlds as well as the physical and its origin.

Divisions of Energy.—And just as the life stream in manifestation flows through differentiated channels of involutionary and evolutionary process, so also does the stream of cosmic energy operate through nine (again a mystic number) as follows:

1	Kinetic Energy	6	Chemical Energy
2	Gravitational Energy	7	Electrical Energy
3	Heat	8	Magnetic Energy
4	Elastic Energy	9	Radiant Energy[49]
5	Cohesive Energy		

All Manifestations of One Source.—At the beginning, however, the Rosicrucian student is taught that all force is a manifestation of the ONE FORCE, all Energy as of the ONE ENERGY and all Substance as ultimately homogeneous.

On a previous page we promised an amplification of the definitions already given, and these we will now take up in order. We will sum up first the POSITIVE ION of which little has been said. We know—

Positive Ions.—1 They are positive electrical conductors, not negative.[50]

2 The velocity of the positive ion is less than that of a corpuscle.

3 Its electrical charge is identical with that of the ordinary atom.

As to Order.

4   Its value of $\dfrac{e}{m}$ is $\dfrac{1}{30,000}$ of $\dfrac{e}{m}$ for a corpuscle.   e=electrical charge. m=mass.

5   Its mass is 1000 times greater than that of a corpuscle and is practically equal to that of an ordinary atom.

6   It can be deflected magnetically, only slightly.

7   From the previous paragraphs it may be seen that as the corpuscle is the connecting link between Ether (of the Spiritual realms) and embryonic Matter (in the Physical worlds) it is the touchstone in the physical search for the "Philosopher's Stone."

It will also be seen that the positive ion has qualities not altogether explained by the corpuscle, and this quality we shall try to arrive at shortly.

Light Rays.—Next come the light rays. The new advances in radio activity have done much to confirm the early teachings of mediaeval Rosicrucians. The "eternal light" is found veritably manifest in the light and fluorescence of radio activity. First of all we will define radio activity itself.

Radio Activity defined.—Radio activity is the explosion of an aggregation of corpuscles comprising unstable, heavy atoms. With the decrease of their kinetic energy, they explode, and the corpuscles rearrange themselves, evolving energy and projecting the products of the rearrangement.[51]

Radio Active substance.—"A radio active substance is one whose atom consists of a complex group of corpuscles, the configuration of which depends for its maintenance upon a certain velocity of movement of the corpuscles comprising it, and beneath which velocity the corpuscles rearrange themselves with the evolution of an amount of energy which breaks down the atom."[52]

Genealogy of Radio Active Light Waves.—Aside from our knowledge of the light waves and rays as exhibited in the spectrum, from red to violet, we have now through the phenomena of radio activity the following rays in their genealogical order:

Cathode Rays

Lenard Rays | X Rays

S Rays | Niewengloski's Rays

N Rays | Becquerel Rays[53]

Transmutation products.—Alpha, Beta and Gamma Rays: These three rays from radium are apparently IDENTICAL WITH

THE POSITIVE IONS, CORPUSCLES AND X RAYS. This is of great importance, for this identity is what makes known the transmutative processes of modern science. For from the radium and other radio active elements are produced NEW or TRANSMUTED ELEMENTS, known to science as URANIUM X[34]

THORIUM X

THORIUM EMANATION[35]

RADIUM EMANATION

RADIUM EMANATION X, 1st, 2nd, 3rd and 4th changes to the final product, and occupying respectively, 22 days, 4 days, 1 minute, 3.7 days, 3 minutes, 21 minutes, 28 minutes, 200 years.

Power and value of Radium.—The radio active power of radium itself is 1,300,000 times that of uranium, with which it is often associated, and one gram would be worth at least $120,000 and a gram is one-twenty-eighth of an ounce. At the present time only a very few grams have been extracted, and radium is obtainable only on the basis of milligrams.[36]

Transmutation a fact.—It is not possible in a limited instruction to go into the actual physical processes employed, but they may be accepted as truth on the status of the eminent men of scientific authority who are sponsors for them. With this in mind let us state that the teaching of the Rosicrucians for centuries that transmutation of elements is possible is borne out by the fact that IT HAS BEEN AND IS BEING DONE NOW, IN THE MODERN LABORATORY. We have three distinct instances:

The Transmuted Elements.—1 The evolution of Uranium X, an entirely NEW Element, out of the decomposition of Uranium. (Rutherford).

2 The evolution of "Exradio" from Radium (Ramsay).

3 The birth of Helium from Radium emanation (Ramsay-Soddy).[37]

Transmutation traditions.—Lower elements, such as Lead and Silver, have been transmuted by exoteric science, and it is a tradition of the Fraternity which seems well attested that Gold has also been raised from lower metals, but the above instances are the actual verifiable results of modern savants, far removed from the allegory, tradition, romance, and fiction of mediaevalism.

Atom compared to a church.—To give an idea of the relative values of sizes, in dealing with infinitesimals, we quote the illustration of Lodge—"If we imagine an ordinary church to be an atom of hydrogen, the corpuscles constituting it would be represented by 100 grains of sand each the size of a period, dashing in all directions inside; rotating with inconceivable velocities, and filling the whole interior of the church with their tumultuous motion. Such an atom would be penetrable to other corpuscles in inverse proportion to the number of corpuscles constituting it, while it would be opaque to other atoms."[58]

Transmutation of Neon and Helium.—One of the latest transmutations are the production of NEON, a gas, and also Helium from bulbs which had contained only Hydrogen. This on the statements of Ramsay and Collie. And note that both Neon and Helium are of higher atomic weight than the Hydrogen from which they appeared.[59]

Number and weight of Atoms.—This has enabled scientists to compute accurately the number of atoms in a given quantity of matter. For instance, a grain of radium gives off 36 billion helium atoms PER SECOND. A cubic centimeter of helium GAS contains 2,560,000,000,000,000,000,000 atoms, while the weight of an atom is 1/68,000,000,000,000,000,000,000,000,000 (octillionths) of a gram. The smallest particle of matter that can be seen with the most powerful microscope contains more atoms than the total number of the human population of the globe since humanity has existed.[60]

"Electricity Is Life."—There is an old saying, "Electricity is Life." Rosicrucians agree with it thoroughly, but go farther. They assert that all that we can know or conceive of is a manifestation of electrical energy, and the statements made in this instruction based upon the findings of authoritative science indicate the activities of both positive and negative electricities in all cosmic operations, and it may not be too much to predicate that the "wisdom of future generations" may find that the "Word that was lost," the Creative Word, is synonymous with electrical activity as a Deific manifestation.

Matter and Electricity identical.—NOW SCIENCE ASSERTS TOO THAT MATTER IS MADE UP OF ELECTRICITY, AND NOTHING BUT ELECTRICITY.[61]

We now know vastly more about electricity than formerly, and instead of it being a hypothetical condition we know WHAT it is in one phase at least. Rosicrucians assert the other phase. What we KNOW about one phase of electricity (the negative) we gain through the Electronic Theory. We will sum it up:[62]

1   Negative electricity consists of unit corpuscles or electrons.

2 Static electricity results from the action of these corpuscles at rest.

3 Current electricity is these corpuscles in motion, whether through gases, liquids or solids.

4. Magnetism is a phase of energy developed in the ether at right angles to the direction of motion of the corpuscles.

5 Light is due to disturbances in the surrounding ether caused by changes in the motion of the corpuscles.

6 Self-induced electricity and mechanical inertia of matter are identical and due to the corpuscle in motion.

7 Mass or matter in quantity is ether carried along by the moving corpuscle; it is not a constant quantity but depends upon the velocity of the moving corpuscle.

8 Atoms are made up of negative charges or corpuscles, each aggregation of corpuscles being surrounded by A SPHERE OF POSITIVE ELECTRICITY. (What IS positive electricity?)

9 THEREFORE, MATTER ULTIMATELY IS IDENTICAL WITH ELECTRICITY.

Positive Electricity the Rosicrucian X Force.—To correlate the spiritual and the physical, the ethical and the material, Rosicrucians teach that POSITIVE ELECTRICITY IS THE "X FORCE," MENTAL POWER AND INTELLIGENCE OF THE ABSOLUTE, which is manifested as the Word and Wisdom of its two Expressions and transmitted via the Radiant Energy and Ether into Creative Process.

The status of the activities of these infinitesimals we have considered defines the status of the planetary bodies of our own and all other solar systems. Metals are not by any means in the same state on all. On the hottest solar bodies we find them in the "proto" state, such as Proto-hydrogen (Pickering of Harvard) in Zeta-Puppis and Argo, and in 29 Canis Major and Gamma-Argus. The element proto-hydrogen is the broken-down element Hydrogen.[63]

Planetary Heat and Radio-Activity.—In the case of our own sun, the comparative youth of our solar system is shown by the fact that our sun cannot have been emitting heat at its present rate more than 18,000,000 years,[64] nor has it illumined this earth for more than 100,000,000 years,[65] even though our earth was thrown off from the sun in the Hyperborean Epoch over 350,000,000 years ago, for the sun did not gain its full power to illuminate until a comparatively recent date, due to the contraction, consequent liberation of heat, and increasing incandescence. *For the heat of all the planets comes from within each, not from the parent sun,* and is due to heat generated by contraction and the radio-activity. Both the sun and our earth contain enormous quantities of radio active matter and radio activity generates a tremendous amount of heat. It is estimated therefore that the radio activity of ordinary substances on both earth and sun will be sufficient to more than offset the amount of heat generated and dissipated into space.[66]

In regard to this matter of heat coming to earth from the sun, a writer in the Electrical Review of January 21st, 1898, presents the truth we have just stated, quite clearly. The article reads:

"It is also assumed that such is the inconceivable effect of combustion in luminous and heat-producing intensity that both light and heat rays are transmitted to the glittering planetary and stellar elements suspended in celestial space, some of which are computed to be hundreds of millions of miles from the solar orb.

"Now anyone who, like the writer, has had great experience in the production of relatively enormous fusion temperatures, will know that although the luminosity resulting from masses of molten metal, such as very low carbon steel, will project a beam of light extending under certain atmospheric conditions, over a distance of five miles; nevertheless the sensible heat transmitted to any body that chanced to intercept such a beam, will not be perceptibly felt at a distance of even, say, 50 yards.

"So that if this ratio of as 50 is to (1,760 x 5) were applied to the proportion of heat and light transmitted from the assumed burning solar orb, no heat sensible to human life, could possibly be transmitted through all the enormous gulf of space that divided our planet from the sun."

Power of Light Pressure, Size of a Particle of Nebulic Mist.— The further power of light is shown by the value of Light Pressure, which is one milligram per square meter of earth's surface, or 70,000 tons for the whole planet, from the sun.[67] This pressure and force exerted on atomic structures causes the continuous changes in the general structural make-up of the earth. The phenomenon of light pressure is illustrated by the force exerted on comets, which when pointing toward the sun have their tails away from it, due to the light pressure bombardment. In this connection the particles of mist which form the comet's tail are analogous to those which form the nebulic mist and are measured as about 1 to 6 thousandths of a millimeter in diameter.

Synthesis of Atoms.—Transmutation is no longer the generally scouted dream of the Rosicrucian alchemist. It is a FACT of 20th Century Science. Sir William Ramsay said, "Experiments are in progress with radio-active substances, the results of which seem to show that we are on the brink of DISCOVERING THE SYNTHESIS OF ATOMS."[68]

What physical scientists seek.—Physical scientists seek to tap the store of inter-elemental energy, "a store so great that every breath we draw has within it sufficient power to drive all the workshops of the world."[69]

What Rosicrucian Initiates seek.—Rosicrucian Initiates, knowing the secret of Positive Electricity, seek to tap the storehouse of Cosmic Conscious Energy, by which life may be prolonged indefinitely, not in a mere mortal shell, but in *conscious sequential memory* through all succeeding incarnations.

Stupendous mathematical evidences cause reverence and humility.—This stupendous range of mathematical evidences of the scale on which Nature and Cosmic forces operate, from the inconceivable dynamic energy displayed in the radio-active bombardments on a microcosmic scale, to the macrocosmic neighborliness of our nearest star Alpha Centauri 26,000,000,000,000 miles distant, with the next nearest twice as far, and the generality at least forty or fifty times as distant, should make us pause in our vaunted conceits of earthly wisdom and bend in humble reverence before works so vast as to evidence beyond all mortal question the manifest intelligence of the Supreme Architect of the Universe, and His assisting Hierarchies.[70]

QUESTIONS ON INSTRUCTION No. 6

1. In what do Occult and Physical Science agree?
2. What is a nebula?
3. What quantities are they known to make up?
4. How are they named?
5. Give examples.
6. What is the "Nebular Hypothesis"?
7. Who formulated and developed it?
8. What does Draper say regarding it?
9. How may nebulae be technically defined?

10. What is the extent of a nebula?
11. What ARE these nebulae?
12. What were the old teachings of physical science?
13. What do the later findings of physical science resolve themselves into?
14. According to this later theory what is a nebula said to be?
15. What are "Planetesimals"?
16. What does the original nucleus of a nebula do?
17. What is the 2nd aspect of a nebula?
18. How does this compare with the Rosicrucian teachings of the creating God?
19. What is the first hypothesis of the origin of a nebula?
20. What is Roche's Law of Limits?
21. What is the teaching of Arrhenius?
22. How do such nebulae become luminous?
23. How is a star born?
24. What does the yellow stage denote?
25. The red stage?
26. How is a cold star rejuvenated?
27. What is cosmic night?
28. How is a star really "cold" at first?
29. What is the joint affirmation of Rosicrucianism and Science?
30. How does Rosicrucianism amplify it?
31. What is said of the "ring theory"?
32. How is axial motion and spherical shape attained?
33. What is the origin of orbital motion?
34. Give a further definition of the origin of nebulae.
35. What is the direction and velocity of our Solar System?
36. What is God said to be?
37. What is one of the greatest cosmic forces?
38. Define "Radiation pressure."
39. What does a nebula consist of wholly or in part?
40. To what does the aggregated matter succumb?
41. What is the shape of our universe?
42. What do all nebulic nuclei pass through?
43. Describe human light rays.

44. On what does light act?

45. What does radiant energy require?

46. What is Mass—Molecule, Atom, Ion, Corpuscles, Electrons?

47. What is the Electronic Theory?

48. What is Ether?

49. What is its family place?

50. What is the Rosicrucian concept of Ether?

51. What is the densest substance known?

52. What is the least substantial?

53. What is the density of Ether?

54. What is Matter composed of principally?

55. What is the size of Ether granules?

56. Where do Rosicrucians and physicists agree?

57. What may Matter be likened unto?

58. What does the universe consist of?

59. What is Nature?

60. What is the creative link?

61. Describe the descent of cosmic substance.

62. What does it make.

63. What is to be the final status of the universe?

64. Describe the action of the 12 around the 13th.

65. Name the attributes of Deity.

66. How do they manifest?

67. What is energy?

68. How many combinations and permutations of the Elements are known?

69. What does Ether furnish?

70. What is the common triad?

71. Name the divisions of energy.

72. Of what are they all manifestations?

73. What are positive Ions?

74. Where may the eternal light be found?

75. Define radio activity.

76. Give its genealogy.

77. Name the transmutation products.

78. How may an atom be compared to a church?

80. What is electricity?

79. Give an idea as to the number and weight of atoms.

81. What do we KNOW regarding electricity and Matter?

82. What is the correlation between positive eletricicty and Rosi-
    crucian X Force?

83. What is learned of planetary heat and radio activity?

84. What of the power of light pressure?

85. What is said of the synthesis of atoms?

86. What do physical scientists seek? Rosicrucian scientists?

87. What do these stupendous mathematical evidences cause?

Printed in the United States
139603LV00001B/8/A